SCIENTIFIC AMERICAN EDUCATIONAL PUBLISHING

LIFE SCIENCE

10 FUN PROJECTS ABOUT BIOLOGY

BRING SCIENCE HOME

Published in 2024 by The Rosen Publishing Group, Inc.
2544 Clinton Street, Buffalo, NY 14224

Contains material from Scientific American, a division of Springer Nature America, Inc., reprinted by permission as well as original material from The Rosen Publishing Group.

First Edition

Editor: Kristen Rajczak Nelson
Designer: Rachel Rising

All illustrations by Continuum Content Solutions

Photo Credits: pp. 4, 5, 8, 10, 13, 15, 18, 20, 25, 26, 33, 36 cve iv/Shutterstock.com; pp. 5, 10, 15, 20, 28, 33, 39, 45, 49, 54 Anna Frajtova/Shutterstock.com.

Activity on p. 5 by Science Buddies (January 19, 2017); p. 10 by Science Buddies (April 3, 2014); p. 15 by Science Buddies (November 17, 2013); p. 20 by Science Buddies (September 19, 2013), p. 28 by Science Buddies (June 13, 2013); p. 33 by Science Buddies (January 31, 2013); p. 39 by Science Buddies (April 19, 2012); p. 45 by Science Buddies (April 16, 2015); p. 49 by Science Buddies (September 3, 2015); p. 54 by Science Buddies (December 13, 2012).

Cataloging-in-Publication Dat
Names: Scientific American, Inc.
Title: Life science: 10 fun projects about biology / edited by the Scientific American Editors.
Description: New York : Scientific American Educational Publishing, an imprint of Rosen Publishing, 2024. | Series: Bring science home | Includes glossary and index.
Identifiers: ISBN 9781725349971 (pbk.) | ISBN 9781725349988 (library bound) | ISBN 9781725349995 (ebook)
Subjects: LCSH: Biology--Experiments--Juvenile literature.
Classification: LCC QH309.2 L54 2024 | DDC 570.72'4--dc23

Manufactured in the United States of America

CPSIA Compliance Information: Batch #CWSA24. For further information contact Rosen Publishing, New York, New York at 1-800-237-9932.

CONTENTS

Introduction 4

How Do Arctic Animals Stay Warm? ⚛ 5

Single-Celled Science: Yeasty Beasties ⚛ 10

Survival Science: Hunting M&M'S Candies ⚛ 15

Soil Science: Make a Winogradsky Column ⚛ 20

Follow Family Traits with an Easy Tree 28

Squishy Science: Extract DNA from Smashed Strawberries ⚛ 33

Hairy Science: Measuring Humidity with a Hair Hygrometer 39

The Big Eggshell Breakdown 45

Cold or Warm, Can We Really Tell? 49

Semisolid Science: Growing Yogurt 54

The Scientific Method 61

Glossary 62

Additional Resources 63

Index 64

⚛ THESE ACTIVITIES INCLUDE SCIENCE FAIR PROJECT IDEAS.

INTRODUCTION

Biology is an important classroom subject at school, but did you know there are fun ways to experiment with biology at home? Whether you're using strawberries, M&M'S, yeast, or yogurt, these 10 projects will teach you cool scientific processes—like fermentation and DNA extraction—as well as encourage you to practice STEM concepts outside of the classroom.

Projects marked with 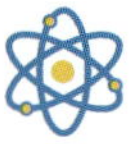 include a section called Science Fair Project ideas. These ideas can help you develop your own original science fair project. Science fair judges tend to reward creative thought and imagination, and it helps if you are really interested in your project. You will also need to follow the scientific method. See page 61 for more information about that.

How Do Arctic Animals Stay Warm?

HOW DO ANIMALS STAY WARM IN SUBZERO TEMPERATURES? TRY THIS COOL ACTIVITY AND SEE WHAT MATERIALS YOU THINK WOULD MAKE THE BEST WINTER LAYERS!

How lovely it is to come home after a chilly winter walk to a cozy house, put on your fluffy slippers and settle by the fireplace with a warming cup of hot chocolate! Animals such as the polar bear, Arctic wolf, or Antarctic penguin are not so lucky to have such a place to keep warm. How do they face the extreme winter temperatures? Most grow "winter coats" in the fall. Did you ever wonder what these coats look like, and how they help keep animals warm? Do this cool activity and you will feel it firsthand!

PROJECT TIME

30 minutes

KEY CONCEPTS

Biology
Physics
Heat transfer
Thermal insulator
Temperature

BACKGROUND

Warm-blooded mammals—including humans—use part of the energy they get from the food they eat to keep their bodily temperatures relatively constant. A lot of energy is needed to regulate your temperature, especially when your environment is way cooler than your body.

Humans use clothes to keep themselves warm—and they need even more layers of clothing the colder it is. Clothes form a thermal barrier between our warm bodies and the cold weather, preventing heat from escaping. Other mammals have developed natural barriers to trap body heat, allowing them to survive in cold climates. It is as if they take on several "winter coats," each of which plays an important part.

The first coat is often a layer of fat hidden under the skin called "blubber." In addition to being a good thermal insulator—it does not allow heat to pass through it easily—fat has the added benefit of serving as a welcome food reserve. The next coat, depending on the animal, is often a dense layer of underfur or down feathers packed closely against the skin like a warm undershirt. This cozy, soft coat is filled with air pockets that provide insulation; a stationary layer of air is another good thermal insulator. Finally, to keep out moisture, there is often a layer of oily, water-repellent guard hairs or feathers. Sometimes these hairs appear white but they are often transparent and hollow, which has a big advantage: they provide extra thermal insulation.

Some Arctic animals, such as polar bears, are so well insulated that they cannot be spotted with night-vision goggles, which pick up radiated heat. This implies their outer layer of fur has the same temperature as their surroundings. In other words, these animals can keep all of their heat inside, losing none to the environment. How amazing!

Ready to feel the insulating capacities of these materials? Gather a few items and try it out!

MATERIALS

- Four zip-top sandwich bags
- Four small plates
- Ice cubes
- Small feathers (Craft feathers are fine.)
- 1/4 pound of butter (one stick) at room temperature
- Ruler

PREPARATION

- Fill one sandwich bag with a layer of feathers about 0.8 inch (2 cm) thick and close the bag.
- Fill another sandwich bag with a 0.8-inch (2 cm) thick layer of butter and close the bag. If the butter was not at room temperature, leave the bag and butter out for awhile before you perform the test.
- Trap air into a third sandwich bag. Try to make it as thick as the other bags. (Closing the bag a little over halfway first, then blowing into it and closing the remainder works well. Because the air you trap this way is warmer than room temperature, it is best to leave the bag out for awhile to adjust.)
- Leave a fourth sandwich bag empty.
- Place a layer of ice cubes on four small plates.
- Place one bag on top of the ice cubes on each plate.

PROCEDURE

- Let one hand rest on the empty bag, the other on the feathered bag. Avoid pressing down on the bags. *Which one feels colder?*
- Remove your hands and wait a moment until both hands feel warm again. Now, place one hand on the empty bag and the other on the bag filled with fat (butter). *Which one feels colder now, the empty bag or the bag with a layer of fat?*
- *Before you go on, which one do you expect to feel colder, the empty bag or the bag with air—or would they both feel as cold?*
- Feel which one feels colder. *Was your prediction correct?*

- In a moment you will compare the layers of feathers, fat, and air. *Which one do you think will feel warmest and which one will feel coldest?* The one that feels warmest creates the best thermal barrier and thus, is the best thermal insulator. Scientists call materials that prevent heat transfer "thermal insulators."
- Perform the test. Make sure to let your hands adjust to room temperature before comparing. *Were you able to tell which bag felt warmer or is the difference too small to detect?*
- For a more detailed test, leave the bags on the ice cubes for 15 minutes and feel again. *Are you able to detect a difference now? Do your predictions and your test results agree?*

EXTRA

Do you think adding several layers will create a bigger thermal barrier? To test this, you first must let your bags warm up again. Take them off the ice and leave them at room temperature for awhile. Then place a stack of two bags on ice cubes and measure how long it takes before the top feels cold. *Does it take longer than with the single layers?*

EXTRA

Try some other materials, such as wool, cotton, leather, fleece, paper, or aluminum foil. *Which ones do you expect to be good insulators? Do your test results agree with your predictions?*

SCIENCE FAIR IDEA

Now that you know some good thermal insulators, could you design a survival suit for a scientist working in the Arctic or a mountain climber exploring Mount Everest? Once you are done with your design, look up what people living or working in extreme cold conditions wear. *Is your design similar or very different? Why do you think they chose their clothing?*

OBSERVATIONS AND RESULTS

The empty bag probably felt a lot cooler than the filled ones. The materials in the bags are good thermal insulators—they do not let heat travel through them easily. Stacking two bags on top of each other creates an even stronger barrier, so it probably took longer before the top bag felt cold.

The bag with butter (a fat) probably felt colder than the bag with air or feathers, leading to the conclusion that heat transfers faster through fat than via an equally thick layer of stationary air. Feathers insulate because they trap air, so there was probably little difference between how the bag of feathers and the bag of air felt.

People in Arctic climates use animal skin and fur as well as synthetic clothes to protect themselves from the cold. Synthetic clothes made to withstand extreme weather use the same heat-trapping techniques as the ones found in animals. For example, they might use down feathers to trap air or have a water-resistant coating to keep out moisture.

CLEANUP

So long as you used a clean bag, you can reuse the butter for cooking. Reuse your feathers for a craft and the other bags for storage. Recycle the bag you blew into.

Single-Celled Science
Yeasty Beasties

TRY OUT A FUNGAL ACTIVITY THAT COMBINES FERMENTATION AND FUN!

Did you know that dry yeast is actually alive? Add the right ingredients, and presto, the mixture becomes a bubbly, oozing mess of life. But just what are the conditions required for this to happen? What does that yeast need to become active and thrive? Try this science activity to find out for yourself!

PROJECT TIME

1 hour

KEY CONCEPTS

Biology
Microorganisms
Microscopic
Metabolism
Carbon dioxide

BACKGROUND

Yeasts are tiny, microscopic organisms—or microorganisms—that are actually a type of fungus. This means that they are more closely related to a mushroom than to plants and animals or bacteria (the latter of which are also microorganisms). These little critters might sound strange and different, but people have been using them for thousands of years to make bread rise. How does this work? It has to do with the metabolism of the yeasts, or, in other words, what they eat and what they turn that food into.

Like us, yeasts must get their food from their surrounding environment to grow and reproduce—that is, to make more yeast. What do they eat? Yeasts feed on sugars and starches, which are abundant in bread dough! They turn this food into energy and release carbon dioxide gas as a result. This process is known as fermentation. The carbon dioxide gas made during fermentation is what makes a slice of bread so soft and spongy. The pockets of gas are produced by yeasts when the dough is allowed to rise before baking.

MATERIALS

- Three plastic 2-liter bottles
- Measuring tablespoon
- White table sugar
- Salt, baking soda, or vinegar
- Permanent marker (optional)
- Measuring cups
- Warm tap water
- One medium-sized pot or bowl, at least 2 quarts (1.9 L) in size
- Six packets of dry yeast or an equivalent amount from a jar
- Three standard-sized latex balloons
- Clock or timer

PREPARATION

- Rinse each bottle thoroughly with water and remove any labels.
- Add 2 tablespoons of sugar to two of the three bottles. *How do you think the sugar will affect the activity of the yeast?*

- To one of the bottles that you added sugar to, also add 2 tablespoon of salt, baking soda, or vinegar. *How do you think adding salt, baking soda, or vinegar will affect the activity of the yeast?*
- Throughout the experiment, keep track of what you added to each bottle. If needed, you can label the bottles with a permanent marker.

PROCEDURE

- Fill the medium-sized pot or bowl with at least 8 cups (1.9 L) of very warm tap water. Adjust the temperature of the hot water coming from the tap until it is almost too hot to hold your hands under. Use this temperature of water to fill the pot.
- Using the warm water from the pot, fill each bottle with about 2.5 cups (600 ml). Put the lid back on to each bottle and shake them each thoroughly to dissolve all of the ingredients.
- To each bottle, add two packets of dry yeast (or an equivalent amount from a jar). Put the lid back on to each bottle and shake each one gently to mix in the yeast.
- Remove each lid and stretch a balloon completely over the opening of the bottle (over all of the ridges). *Why do you think it is important to form a tight seal with the balloon on the bottle's opening?*
- Leave the bottles to rest in a warm location for 45 minutes. Keep the balloons out of direct sunlight. *How do the balloons change over time?*
- After 45 minutes, examine the bottles and the balloons. *Which balloons have become inflated? How big are they compared to each other? Do you notice any differences in the contents of the bottles?*
- *In which environment did the yeast make the most carbon dioxide? What does this tell you about the conditions needed for yeast fermentation to take place?*

EXTRA

You could quantify your results from this activity by using a water displacement test. To do this, you could fill a large pot completely full with water, place it in a larger tray, pan, or pot. Quickly tie off the balloon you would like to measure without letting gas escape, and then submerge the balloon in the water. You can measure how much water overflowed from the pot into the tray to determine how much water the balloon displaced, and consequently the volume of the carbon dioxide gas inside the balloon. *If you quantify your results, exactly how different are the sizes of the balloons?*

SCIENCE FAIR IDEA

Another environmental condition that can affect the activity of yeast and the process of fermentation is temperature. You could explore this by preparing several bottles using the same conditions, and then placing each bottle in a different place with a different temperature. *After 45 minutes, how do the balloons vary in size?*

SCIENCE FAIR IDEA

You could try this activity again, but next time focus on how using different types and sources of sugars affect the carbon dioxide production. *How do the sugars from different juices or other sources affect how much carbon dioxide is produced?*

OBSERVATIONS AND RESULTS

Did the balloon on the bottle with only yeast and water remain uninflated? Did the balloon on the bottle with only sugar added inflate the most?

When yeasts eat sugar and turn it into energy, they also produce carbon dioxide. This process is known as fermentation. In this activity, the balloons on the bottles should have captured carbon dioxide produced by the yeasts during fermentation. In the bottle that contained yeasts but not sugar, the yeasts did not have food (i.e., sugar) so the balloon should not have inflated. In the bottle that contained yeasts and sugar (but not salt, baking soda, or vinegar), the yeasts should have thrived and made a lot of carbon dioxide, clearly inflating the balloon. When salt, baking soda, or vinegar was added, the yeasts should have made less carbon dioxide, inflating the balloon less than when only sugar was used. This is because the addition of these substances changed the environment and made it less ideal for the yeasts. Specifically, adding salt increased the salinity of the environment, and adding baking soda or vinegar changed the pH of the environment, making it more basic or acidic, respectively, compared to the neutral environment provided by the plain water.

CLEANUP

When you are done with this activity, dispose of the yeasts by composting them or (with permission) dumping them outside somewhere. Do not pour the yeasts down the drain without diluting them with water, as they may damage pipes when they expand.

Survival Science
Hunting M&M'S Candies

THIS CANDY CAMOUFLAGE CHALLENGE IS SURE TO BE AS SWEET AS IT IS EDUCATIONAL.

Have you ever wondered how predators such as wolves, lions, and hawks are able to find their prey? And what can an animal do to keep from being a predator's dinner? To survive, some animals use camouflage so they can better blend in with their surroundings. In this science activity, you'll be the hungry predator and you'll hunt for different colored candies. But it may not be as easy as it sounds—some of your prey will be camouflaged by their habitats. Will they be able to avoid your grasp? To find out, work up an appetite and go hunting for leftover candy!

PROJECT TIME

40 minutes

KEY CONCEPTS

Animals
Predators
Prey
Camouflage

BACKGROUND

Nature can be brutal. In the animal world, if you're not a hunter, then you're likely being hunted. (And sometimes you're both the hunter and the hunted.) What do animals do to avoid being eaten? Some animals develop defense mechanisms such as porcupine quills, armadillo armor, gross tastes on their bodies, or even poisons. This pattern of animals developing strategies to survive occurs thanks to adaptation—and it is a mechanism of evolution.

A common way that animals can avoid being eaten by a predator is by using an adaptation called camouflage, which is a set of colorings or markings on an animal that help it to blend in with its surroundings, or habitat, and prevent it from being recognized as potential prey. This increases its chance for survival. Different animals need to adapt to different habitats, and animals adapt in all kinds of interesting ways using camouflage. For example, katydids are insects that usually live in leafy trees, and many have adapted to that habitat by developing a green body shaped like a leaf. This helps them hide from predators such as birds, who might mistake them for less tasty leaves. Some species of octopuses often change habitats on colorful coral reefs and have adapted a way of changing their skin to match their immediate environment.

MATERIALS

- Six plastic bags
- Plain M&M'S, at least 10 of each color
- Skittles, at least 60 of each color
- Metal pie tin or sturdy paper plate
- Timer or stopwatch
- Two to four volunteer "predators" who are up for devouring some M&M'S

PREPARATION

- Count and place 10 M&M'S of each color into a plastic bag. This means you should have one plastic bag with 10 yellow, 10 blue, 10 green, 10 brown, 10 red, and 10 orange M&M'S candies in it (making a total of 60 candies in the bag).

- Count and place 60 Skittles of each color into their own bags (making five separate bags each containing a separate color). This means you should have one plastic bag with 60 orange Skittles, one with 60 yellow candies, one with 60 green candies, one with 60 red candies, and one with 60 purple Skittles. If you are short on time, you can skip preparing one or two of these bags.

PROCEDURE

- Explain to your two to four volunteer "predators" that they should pretend to be hungry M&M'S-feasting birds. They should make a "beak" using their pointer finger and thumb for collecting M&M'S. Explain that they'll have 20 seconds to use their beaks to quickly pick up M&M'S and put them in their other hand. To encourage the volunteers to be fast, tell them that when they are done with the activity, they can eat the same number of candies as they picked up. But they should not eat the candies until you are done with the activity.

- Also tell the volunteers that they should avoid picking up any Skittles candies, because Skittles make the M&M'S birds sick. The M&M'S are their prey, and the Skittles represent the habitat in which the M&M'S live—and try to use as camouflage. *How do you think the Skittles habitat will work to camouflage the different colored M&M'S prey?*

- After explaining these rules, pour one prepared bag of Skittles into a metal pie tin or sturdy paper plate. Mix in the prepared bag of M&M'S. Put the pie tin in the middle of your group of M&M'S bird volunteers. Make sure everyone can reach the pie tin. *Which M&M'S are the best camouflaged in your pie tin?*

- Set your timer for 20 seconds. Say "Go!" and start the timer. When the timer beeps, make sure everyone stops picking up M&M'S.

- Count the number of each M&M'S color that each person collected. Also count any Skittles that were picked up. *Which M&M'S color was the least-picked one? What do you think this has to do with camouflage?*

- Put all of the M&M'S back in the bag you prepared them in (including M&M'S that people picked). Remove the Skittles you used for the habitat (by pouring them off the pie tin).
- Repeat the 20-second M&M'S hunt with the other prepared bags of Skittles until you have tested each Skittles habitat (separately) with the M&M'S. *For each Skittles habitat, which M&M'S color was the least-picked one? Can you explain what this has to do with camouflage?*

SCIENCE FAIR IDEA

Adaptation and evolution happen over several generations. Try testing this by picking one colored Skittles habitat and mixing your prepared bag of M&M'S with it in the pie tin. After a 20-second round of M&M'S predation, as described in this activity, double the number of M&M'S that are left by adding a colored M&M'S to match the color of each remaining M&M'S. For instance, if there are four red and two brown M&M'S, then add four more red and two more brown M&M'S. Then repeat another round of predation. *How many rounds does it take to remove all of the M&M'S of one color?*

EXTRA

Find pictures of different camouflaged animals, such as a katydid that looks like a leaf on a tree branch or an octopus that is blending in with its surroundings. After finding several pictures of different camouflaged animals, show them to some volunteers and time how long it takes the volunteers to spot each animal. *Do some camouflage techniques work better than others, causing the volunteers to take longer to spot the animal?*

EXTRA

You could try repeating this activity using different types of candy. *How easy is it for M&M'S birds to catch their prey when the habitat is made using candies of a different shape, such as Nerds or other candies that are both different in shape and multicolored, like candy corns?*

OBSERVATIONS AND RESULTS

Did the M&M'S that were the same color as their habitat get picked less frequently? Did the brown M&M'S get picked less frequently in the purple Skittles habitat than any other M&M'S color?

At a glance, it is difficult to tell M&M'S and Skittles apart. Skittles are slightly different in shape from M&M'S, but their yellow, green, red, and orange colors are similar. So if you tell someone to avoid picking Skittles (the habitat color), they're probably less likely to pick M&M'S that are the same color as those surrounding Skittles. Those M&M'S "prey" are being camouflaged by the same-colored (and similarly shaped) Skittles habitat. Consequently, in the yellow Skittles habitat, the yellow M&M'S should have been picked less often than any other variety, and the same goes for the green, red, and orange Skittles habitats—the same-colored M&M'S should have been picked the least often. Also, purple Skittles are somewhat similar in color to brown M&M'S, so the brown ones should have been the color least often picked in the purple Skittles habitat. Blue M&M'S should have been picked at about the same rate in each habitat because there are no Skittles that are very close in color.

CLEANUP

When you are done with your experiment, you can let the volunteers eat the candies if they want to.

Soil Science

Make a Winogradsky Column

WITH THIS MICROBE-DEPENDENT DEMONSTRATION, THE COLUMNS YOU CREATE WILL SHOW WHICH NUTRIENTS ARE HELPFUL OR HARMFUL TO SOIL.

PROJECT TIME

1.5 hours plus 6 to 8 weeks of observation time

KEY CONCEPTS

Microorganisms
Soil
Biology
Oxygen
Sunlight
Ecosystem

Our planet recycles and reuses everything on it needed to support life. It's an amazing, giant recycling system called the biogeochemical cycle. You can actually model this on a small scale by using a plastic bottle and mud to build what's called a Winogradsky column. In this activity, you'll build your own Winogradsky columns and investigate how including different nutrients affects which soil microorganisms flourish and which fail.

BACKGROUND

To live and grow, organisms need specific nutrients nearby for them to eat, just like we do. Most of these nutrients are continuously being moved through a biogeochemical cycle, which transports nutrients and other chemicals through the living and nonliving parts of Earth—for example, iron that was once in a rock might later enter the soil where it can be absorbed by a plant and then eaten by an animal. The biogeochemical cycle is a closed system, which means that the nutrients aren't lost or created; they're continuously reused and recycled.

A given ecosystem might have several biogeochemical cycles going on within it. Some important nutrients that are recycled through an ecosystem are oxygen, carbon, and sulfur. Different soil microbes play key roles in recycling these and other nutrients. A tool that microbiologist Sergei Winogradsky invented to study these processes was a long, sealed column of muddy soil, now called a Winogradsky column. Within one of these, different gradients form. For example, over time there is more oxygen at the top of the column than at the bottom. These gradients affect where different microbes can live within the column.

MATERIALS

- Four clean plastic water bottles that hold 2 cups (473 ml) each. Bottles that have smooth sides (not ridges) and are taller and narrower in shape should work best.
- Scissors
- Knife (optional)
- Permanent marker
- Ruler
- Rubber boots and old clothes that can get muddy (optional)
- Access to a muddy stream, pond, lake, or marsh
- Gloves
- Shovel or trowel
- Two buckets
- Newspaper or plain paper (shredded)
- An egg
- Two bowls
- Two large mixing bowls
- Measuring cup
- Wide stick, for packing mud into the bottles
- Measuring teaspoon
- Adjustable desk lamp with 13-watt compact fluorescent bulb (optional)

- An empty surface, such as on a desk or table, that is at warm room temperature (about 72 to 78°F [22.2 to 25.6°C]). It should not receive any direct sunlight but if you're not using a desk lamp, the area should be near a very sunny window.
- Plastic wrap
- Plastic trash bags or grocery bags
- Four rubber bands
- Cardboard box or brown paper bag
- Flashlight

PREPARATION

- Remove any wrappers from your four water bottles as best you can.
- Carefully cut the tops off your bottles. You may need to use scissors and a knife to do this, and an adult might need to help. Cut the top off right where the bottle starts curving inward near the top. Save the cut-off tops to use as funnels later.
- Using a permanent marker, make a small mark where your bottle would be about 85 percent full. For example, if your bottle is 6 inches (15 cm) tall, you would make a mark at about 5 inches (13 cm) up from the bottom. You will fill the bottles to this level with mud.
- Label each bottle with what will be added to its mud. One will have newspaper added, one will have egg yolk added, and two will only hold plain mud (with nothing added). One of the plain bottles will be kept in the dark, the other three bottles will be exposed to light.
- Go collect some mud from a muddy stream, pond, lake, or marsh. Use caution around the water and always have adult supervision. (You may want to first put on some rubber boots and old clothes that can get muddy.) Put on gloves and fill one bucket with about 1/2 to 1 gallon (1.8 to 3.7 L) of mud, scooping it mostly from just below the water's surface.

- In the other bucket, collect some water (1/2 gallon [2 L] at most) from the same location.
- Bring your buckets home and go wash up if needed! (Until you use them, keep your buckets shaded so they don't get too hot.)

PROCEDURE

- In a bowl, cut about a quarter sheet of newspaper or plain paper into thin strips, and then cut the strips into tiny rectangles. These will be a source of carbon for the microbes in the mud. *How do you think adding carbon will affect the microbes that grow in the Winogradsky column?*
- In a second bowl, add an egg yolk (raw or hard-boiled). If it is hard-boiled, crush the yolk into a mash. If it is raw, be sure to wash your hands after handling it, as raw eggs may contain salmonella bacteria. The yolk will be a source of sulfur for the microbes in the mud. *How do you think adding sulfur will affect the microbes that grow in a Winogradsky column?*
- Put on some gloves and get your bucket with your mud. While stirring, slowly add the water you collected to the soil until your mixture becomes like a milk shake. Also while stirring, be sure to pick out all sticks, leaves, and rocks from the mud.
- In a large mixing bowl, blend a little more than 2 cups of the mud and about one-third of the egg yolk. Using a cut-off bottle top as a funnel, pour about 1 inch (2.5 cm) of the mud mixed with egg yolk into the bottle labeled as having egg yolk added. Tap the bottle on a hard surface to pack the mud down and use a wide stick to pack the mud more. Continue adding about 1 inch (2.5 cm) of mud at a time and packing it down until you are at the 85 percent full mark you made. Set the bottle aside.

- In another large mixing bowl, blend a little more than 2 cups of the mud from the bucket (remix if water has settled out) and 1 teaspoon of the shredded newspaper or paper. Using a cut-off bottle top as a funnel, pour and pack about 1 inch (2.5 cm) of the mud at a time into the correctly labeled bottle until you are at the 85 percent mark. Set the bottle aside.

- Take the two remaining bottles, which you should have labeled as having plain mud. Into the two bottles, funnel and pack in mud directly from the bucket of mud (remix if water has settled out) until they're 85 percent full.

- After sitting about 30 minutes, the water should be about 0.2 to 0.8 inch (0.5 cm to 2 cm) deep on top of the mud in each bottle. Carefully add more water or remove some, as needed. Leave at least 0.2 inch (0.5 cm) of empty space at the top.

- Find an empty flat surface, such as a desk or tabletop, that is at warm room temperature (about 72 to 78°F [22.2 to 25.6°C]). It should not receive any direct sunlight. If you're not using a desk lamp, the area should be near a very sunny window. Cover the surface with plastic trash or grocery bags to protect it.

- Carefully move your bottles to the surface, being careful not to spill them! Cover each bottle with plastic wrap, secured with rubber bands. Your Winogradsky columns are now ready for testing!

- If you're not using a desk lamp, arrange the three Winogradsky columns (that will receive light) so that they'll receive a lot of light, but they should not be exposed to direct light. If you're using a lamp, arrange the three columns so that they're 20 inches (51 cm) away from the lightbulb. For those bottles that are being lit by lamp, write a small "L" on the sides of the bottles that are facing the lamp. *What do you think will happen on the sides facing the light?*

- Set the Winogradsky column with plain mud that won't receive light on a surface at room temperature. Then either place a cardboard box upside down over the column, or put it in a brown paper bag, so that no light reaches the bottle. *What do you think will happen in the column that doesn't receive light?*

- For the next six to eight weeks, leave the Winogradsky columns where you set them. If you're using one, keep the desk lamp on 24 hours a day. (If a plastic wrap lid comes loose, just reattach it with a rubber band.) Observe the columns once a week, looking for color changes in them. Each area of coloring should be a group of the same type of microbes. When observing the columns, try turning off the lights and shining a bright flashlight at the columns—this can help you see the colors better. *What colors do you see appear in the columns? Where in the columns do they appear? How do the columns look different from one another? Do you see any worms, shrimp, snails, or other larger organisms in the columns? How do the columns change over time? What do you think your results have to do with what was added (or not added) to the columns?*

- **Tip:** You should see some green coloring appear on some columns after one to two weeks. You may need to look closely for it. If you do not see any green coloring, the columns may not be receiving enough light. You could try moving them closer to the light source.

SCIENCE FAIR IDEA

Do this activity again, but this time test multiple bottles for each condition. For example, test three Winogradsky columns with egg yolk. *How reproducible are your results? Is there a lot of variation between the different columns that were set up the same way?*

SCIENCE FAIR IDEA

You could try testing several different sources of mud or soil to see if the microbial growth will be different from location to location. You could even try some beach sand. *What do you think your results tell you about the soil quality and microbes that live at each site you test?*

SCIENCE FAIR IDEA

Test some different kinds of additives to look for microbes that live in unique and challenging environments. You could test increasing amounts of salt in a series of Winogradsky columns to test for salt lovers (called halophiles) or place columns at different temperatures to find microbes that like heat (near a heat vent) or cold (in the refrigerator). *Can you select for microbes that live in more extreme conditions?*

OBSERVATIONS AND RESULTS

Did the columns that were in the light make areas of green coloring on the sides facing the light whereas the column in the dark remained dark brown? Did the three columns that were in the light create color patterns that were somewhat different from one another?

Over time, gradients of different nutrients should have formed in the Winogradsky columns. These gradients affect where different microbes grow within the columns. For example, over time there's more oxygen at the top of a column than at the bottom, and this means that microbes that can tolerate or make oxygen will be at the top. Microbes that cannot tolerate free oxygen (called anaerobic bacteria) will be further down. Similarly, microbes that need light to make energy (via photosynthesis or a similar process) will need to live where they can get light in the column.

After about one to two weeks, depending on how much light the columns receive, some green coloring should appear in the columns receiving light on the illuminated sides. This is mostly due to cyanobacteria and algae, which need light. The column in the dark should remain dark brown. In the column that had egg yolk you may have seen areas of darker green, purple, and/or black coloring develop over time near the bottom—these colorings could be groups of certain anaerobic bacteria: green sulfur bacteria, purple sulfur bacteria, and sulfate-reducing bacteria, respectively. Sulfate-reducing bacteria actually eat sulfur and make hydrogen sulfide gas, which is eaten by the green and purple sulfur bacteria. In the column that had newspaper, you may have seen some areas of brown, orange, red, or purple near the middle—these colorings could be groups of purple nonsulfur bacteria, which need a carbon source to thrive. You may have seen worms, snails, shrimp, or other small organisms in the water, but probably not many (if any) in the bottle with the egg yolk because hydrogen sulfide is toxic to most organisms!

CLEANUP

Be sure to wash your hands and anything else that came into contact with the raw egg, because eggs can carry salmonella. Also wash your hands after handling the mud. When you are done with your Winogradsky columns, with permission you can dump the mud outside (such as in a composter or back in a muddy area).

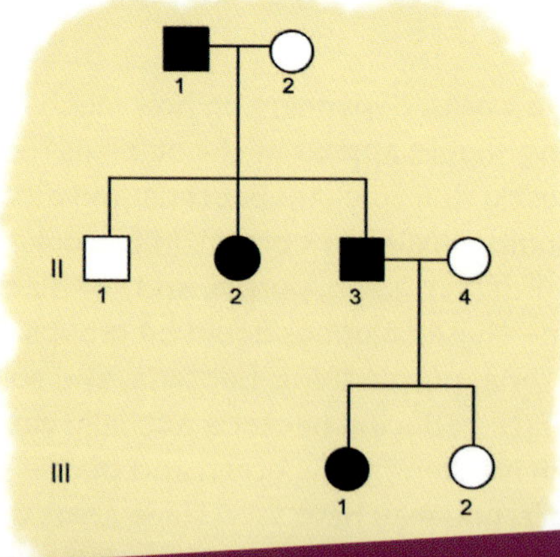

Follow Family Traits with an Easy Tree

FIGURING OUT YOUR FAMILY TRAITS IN THIS ACTIVITY WILL HAVE YOU DESTINED TO DRAW SIMILARITIES!

Has anyone ever told you that you look just like one of your parents or grandparents? Some characteristics, such as the shape of your hairline or whether your earlobes are attached or detached, are inherited. In this activity, you'll get to see how writing some characteristics onto a family tree can help you determine just how you inherited them.

PROJECT TIME

25 minutes

KEY CONCEPTS

Genetics
Inheritance
Family
Physical traits

BACKGROUND

When we look at members of a family, it's easy to see that some physical characteristics, or traits, are shared. In the 1860s, botanist Gregor Mendel discovered that in pea plants some physical features, such as flower color, are passed down in clear and predictable patterns. Today we know that offspring inherit their DNA from both parents—half from each. This results in two copies of every gene, which is composed of a variety of pairs of DNA. Many genes come in different versions, called alleles. Alleles are differences in the DNA sequence of a gene. If you have two identical alleles, you are said to be homozygous for that gene whereas if you have two different alleles you're heterozygous.

Some alleles are dominant, meaning that if you have one copy of that allele you will display that trait. Other alleles are recessive, meaning you need two copies of that allele to display the trait. For example, Mendel took pea plants that were homozygous for different traits and crossbred them. When crossing homozygous purple flowered plants to homozygous white flowered plants, the offspring (which were heterozygous) had purple flowers. The purple allele was dominant and the white one was recessive. When the heterozygous purple-flowered offspring were crossed with one another, some of their offspring would wind up with two copies of the recessive allele, giving them white flowers.

MATERIALS

- Paper and a pencil or pen
- Four sheets of paper
- At least three generations of people from the same family. The more members of the family that are available, the better the results will be.

PREPARATION

- Draw a family tree, or pedigree, showing the different members of the family. Include all of the family members from whom you will be getting data. You can designate the males by a square and the females by a circle. You can look at this resource on "Your Family Health History" (https://www.genome.gov/Pages/Education/Modules/YourFamilyHealthHistory.pdf) for examples of family trees.

- Make three copies of the family tree (so you have four total).
- Label each family tree one of the following: "Earlobes," "Widow's Peak," "Mid-Digit Hair," and "Hitchhiker's Thumb."

PROCEDURE

- Gather the family members together that you put on the family tree. Alternatively, you can do this activity with some members separately.
- Determine whether each family member has attached or detached earlobes. (Attached earlobes are clearly attached to the side of the head at the bottom of the earlobe whereas detached ones hang free.) There can be a range of attachment—just do your best to determine how attached the earlobes are. *Do many relatives have the same type of earlobe or is there variation? Can you see how this trait was inherited?* Write down your results in your "Earlobes" family tree.
- Determine whether each family member has a widow's peak or not. (A widow's peak is where the hairline comes to a V-shaped point above the person's forehead.) Keep in mind that widow's peaks can vary considerably—when determining if a person has a widow's peak, count any sort of V-shaped hairline as a widow's peak. *Do many relatives either have a widow's peak or don't have one? Can you see how this trait was inherited?* Write your results on your "Widow's Peak" family tree.
- Determine whether each family member has hair on their mid-digits (the middle joints of the fingers) or does not have it. You may need to look closely at each person's hands—if they have any hair on the mid-digit, even a strand, then they have mid-digit hair. *Do many relatives have hair on their mid-digits or are they hairless? Can you see how this trait was inherited?* Write your results on your "Mid-Digit Hair" family tree.

- Have each family member make a fist with their thumb sticking up. Determine whether their thumb when extended is straight or curved. (A curved thumb is also called a hitchhiker's thumb.) Keep in mind that thumbs come in a wide range of curvedness, from completely straight to completely curved—do your best to decide if a person's thumb looks curved or straight. *Do many relatives have a hitchhiker's thumb or are their thumbs mostly straight? Can you see how this trait was inherited?* Write your results in your "Hitchhiker's Thumb" family tree.

- *Overall, can you see how different traits were passed down through your family? Are there traits that skipped a generation? Are there traits that were in every generation? Can you figure out which traits might be mostly recessive and which might be mostly dominant?*

EXTRA

There are several other genetic traits that you could investigate using this activity. *How does it look like other traits are inherited, such as freckles, cleft chins, toe lengths (whether the big or the second toe is longest), and which thumb is on top when you interlace your fingers?*

EXTRA

In this activity, you looked at existing family members to investigate how different traits are inherited. *Can you use your knowledge from this activity to predict what traits future offspring of different family members might have if you know what traits their partner has?*

EXTRA

People have long debated whether handedness is a genetic trait. One scientific study showed a correlation between handedness and hair whorl direction, specifically suggesting that a greater percentage of left-handed people have counterclockwise hair whorls than do right-handed people. *Can you find a correlation between handedness and the direction of peoples' hair whorls in your family?*

OBSERVATIONS AND RESULTS

Could you see how some traits were passed down? Did it look like having detached earlobes, a widow's peak, mid-digit hair, and a straight thumb were all dominant traits?

Although the presence or absence of attached earlobes, a widow's peak, mid-digit hair, and a hitchhiker's thumb are thought to be primarily genetically inherited traits, there is some controversy over exactly how they are inherited. In other words, they may be affected by more than one gene, more than two alleles, or factors other than genetics. That said, having detached earlobes, a widow's peak, mid-digit hair, and a straight thumb are generally considered to be dominant traits, although clearly their inheritance is complex. A dominant trait is one that only needs one copy of the dominant allele to be displayed. For a person to show a recessive trait, they generally need two alleles for the recessive trait. Because a person only needs one copy of the dominant allele to show the trait, they could be homozygous or heterozygous for the allele. This means that a person who shows a dominant trait could have a child with a person who shows the recessive trait (or even another person who shows the dominant trait) and the child may show either trait (because the parent with the dominant trait may be heterozygous). Two people who both show a recessive trait are most likely to have a child that also shows the recessive trait, but inheritance is often affected by other factors.

CLEANUP

Share the family trees with your volunteers.

Squishy Science

Extract DNA from Smashed Strawberries

A GENETICALLY GEARED ACTIVITY THAT WILL REQUIRE YOU TO EXTRACT SOME KNOWLEDGE.

Have you ever wondered how scientists extract DNA from an organism? All living organisms have DNA, which is short for deoxyribonucleic acid; it is basically the blueprint for everything that happens inside an organism's cells. Overall, DNA tells an organism how to develop and function. DNA is so important that this complex compound is found in virtually every one of an organism's cells. In this activity, you'll make your own DNA extraction kit from household chemicals and use it to separate DNA from strawberries.

PROJECT TIME

35 minutes

KEY CONCEPTS

DNA
Genome
Genes
Extraction
Laboratory techniques

BACKGROUND

Whether you're a human, rat, tomato, or bacterium, each of your cells will have DNA inside of it (with some rare exceptions, such as mature red blood cells in humans). Each cell has an entire copy of the same set of instructions, and this set is called the genome. Scientists study DNA for many reasons: They can figure out how the instructions stored in DNA help your body to function properly. They can use DNA to make new medicines or genetically modify crops to be resistant to insects. They can solve who is a suspect of a crime, and can even use ancient DNA to reconstruct evolutionary histories!

To get the DNA from a cell, scientists typically rely on one of many DNA extraction kits available from biotechnology companies. During a DNA extraction, a detergent will cause the cell to pop open, or lyse, so that the DNA is released into solution. Then alcohol added to the solution causes the DNA to precipitate out. In this activity, strawberries will be used because each strawberry cell has eight copies of the genome, giving them a lot of DNA per cell. (Most organisms only have one genome copy per cell.)

MATERIALS

- Rubbing alcohol
- Measuring cup
- Measuring spoons
- Salt
- Water
- Dishwashing liquid (for hand-washing dishes)
- Glass or small bowl
- Cheesecloth
- Funnel
- Tall drinking glass
- Three strawberries
- Resealable plastic sandwich bag
- Small glass jar (such as a spice or baby food jar)
- Bamboo skewer, available at most grocery stores. (If you use a baby food or short spice jar, you could substitute a toothpick for the skewer.)

PREPARATION

- Chill the rubbing alcohol in the freezer. (You'll need it later.)
- Mix 1/2 teaspoon of salt, 1/3 cup (79 ml) of water, and 1 tablespoon (15 ml) of dishwashing liquid in a glass or small bowl. Set the mixture aside. This is your extraction liquid. *Why do you think there is detergent in the extraction liquid?*
- Completely line the funnel with cheesecloth. Insert the funnel tube into the tall drinking glass (not the glass with the extraction liquid in it).
- Remove and discard the green tops from the strawberries.

PROCEDURE

- Put the strawberries into a resealable plastic sandwich bag and push out all of the extra air. Seal the bag tightly.
- With your fingers, squeeze and smash the strawberries for two minutes. *How do the smashed strawberries look?*
- Add 3 tablespoons (44 ml) of the extraction liquid you prepared to the strawberries in the bag. Push out all of the extra air and reseal the bag. *How do you think the detergent and salt will affect the strawberry cells?*
- Squeeze the strawberry mixture with your fingers for one minute. *How do the smashed strawberries look now?*
- Pour the strawberry mixture from the bag into the funnel. Let it drip through the cheesecloth and into the tall glass until there is very little liquid left in the funnel (only wet pulp remains). *How does the filtered strawberry liquid look?*
- Pour the filtered strawberry liquid from the tall glass into the small glass jar so that the jar is one-quarter full.
- Measure out 1/2 cup (118 ml) of cold rubbing alcohol.

- Tilt the jar and very slowly pour the alcohol down its side. Pour until the alcohol has formed approximately a 1-inch (2.5 cm) deep layer on top of the strawberry liquid. You may not need all of the 1/2 cup (118 ml) of alcohol to form the 1-inch (2.5 cm) layer. Do not let the strawberry liquid and alcohol mix.

- Study the mixture inside of the jar. The strawberry DNA will appear as gooey clear/white stringy stuff. *Do you see anything in the jar that might be strawberry DNA? If so, where in the jar is it?*

- Dip the bamboo skewer into the jar where the strawberry liquid and alcohol layers meet and then pull up the skewer. *Did you see anything stick to the skewer that might be DNA? Can you spool any DNA onto the skewer?*

SCIENCE FAIR IDEA

You can try using this DNA extraction activity on lots of other things. Grab some oatmeal or kiwis from the kitchen and try it again! *Which foods give you the most DNA?*

EXTRA

If you have a milligram scale (called a balance), you can measure how much DNA you get (called a yield). Just weigh your clean bamboo skewer and then weigh the skewer again after you have used it to fish out as much DNA as you could from your strawberry DNA extraction. Subtract the initial weight of the skewer from its weight with the DNA to get your final yield of DNA. *What was the weight of your DNA yield?*

EXTRA

Try to tweak different variables in this activity to see how you could change your strawberry DNA yield. For example, you could try starting with different amounts of strawberries, using different detergents, or different DNA sources (such as oatmeal or kiwis). *Which conditions give you the best DNA yield?*

OBSERVATIONS AND RESULTS

Were you able to see DNA in the small jar when you added the cold rubbing alcohol? Was the DNA mostly in the layer with the alcohol and between the layers of alcohol and strawberry liquid?

When you added the salt and detergent mixture to the smashed strawberries, the detergent helped lyse (pop open) the strawberry cells, releasing the DNA into solution, whereas the salt helped create an environment where the different DNA strands could gather and clump, making it easier for you to see them. (When you added the salt and detergent mixture, you probably mostly just saw more bubbles form in the bag because of the detergent.) After you added the cold rubbing alcohol to the filtered strawberry liquid, the alcohol should have precipitated the DNA out of the liquid while the rest of the liquid remained in solution. You should have seen the white/clear gooey DNA strands in the alcohol layer as well as between the two layers. A single strand of DNA is extremely tiny—too tiny to see with the naked eye—but because the DNA clumped in this activity, you were able to see just how much of it three strawberries have when all of their octoploid cells are combined! ("Octoploid" means they have eight genomes.)

CLEANUP

Safely dispose of the DNA extractions and wash the kitchen supplies used.

Hairy Science

Measuring Humidity with a Hair Hygrometer

THIS IS A RAINY DAY ACTIVITY THAT MIGHT LEAVE YOU FEELING HAIR-IED!

Does your hair go crazy during "April showers," when the weather turns damp? If so, have you wondered why this is? Strands of hair can relax and lengthen when the humidity increases, and then contract again when the humidity decreases. In fact, the rate of change in the length of hair strands is so dependable that they can actually be used as the basis for a hygrometer, a device that measures the humidity level in the air.

PROJECT TIME

40 minutes

KEY CONCEPTS

Weather
Humidity
Biology
Chemistry

BACKGROUND

To understand how a hygrometer made with hair strands works, it is important to understand the structure of a shaft of human hair. (The shaft is the part of the hair that sticks out above the surface of the skin, whereas the hair follicle is the part that is below the skin's surface.) In human hair shafts, several outer layers of flattened cuticle cells surround a layer of cortical cells (making up the cortex), which surrounds the central core of cells, called the medulla. The flattened cuticle cells are relatively tough and have a scalelike appearance when magnified, along with a coating of specialized molecules (called lipids) that repels water, helping to protect the hair strand. The cortical cells are fibrous and contain pigment granules of varying darkness that give hair strands their natural color.

The predominant proteins in hair are from a group called keratins, from the same protein family that forms your fingernails. The chemical bonds that link the keratin molecules are what give a hair strand its strength and flexibility as well as its predictable response to humidity.

MATERIALS

- Rubbing alcohol (70 percent isopropyl alcohol)
- Measuring spoons
- Small bowl
- Three long strands of human hair (about 8 inches [20 cm] long)
- Cotton swabs
- Measuring tape or ruler
- Thin, flat piece of plastic that can easily be cut (about 3 inches long by 3 inches wide [8 cm long by 8 cm wide]), such as from the lid of a disposable deli container
- Scissors (strong enough to cut the plastic piece)
- Dime
- Tape
- Two small nails
- Glue
- Scrap piece of wood or flat Styrofoam (about 10 inches long and 4 inches wide [25 cm long by 10 cm wide])
- Hammer
- Pencil or pen
- Hair dryer

- Plastic box that can be sealed and is large enough to fit the wood or Styrofoam piece inside of it
- Wet sponge or small towel (either paper or cloth)

PREPARATION

- Make a solution of 25 percent rubbing alcohol and 75 percent water by mixing 1 tablespoon (15 ml) of water with 1 teaspoon (5 ml) of rubbing alcohol in a small bowl.
- Dab a cotton swab in the diluted rubbing alcohol solution and wipe down each hair strand. This will remove any residual oils and allow water to permeate the hair strands more easily. Set the hair strands aside somewhere where you won't lose them.
- Cut the piece of plastic into a triangular pointer shape, specifically an isosceles triangle that is 2.8 inches (7 cm) on two sides and 1.2 inches (3 cm) on the third side.
- Position the plastic pointer you made so that it is pointing to your right, then tape a dime onto the pointer, centered 1.8 inches (4.6 cm) away from the pointer's short (left) side. (Some of the dime will stick out a little above and a little below the other two sides of the pointer.)
- Centered about 0.2 inch (0.5 cm) from the left side of the pointer, poke one of the nails through it. Be careful when using the hammer on the nails. Be sure to only use the hammer on a surface that can safely withstand the force and is resistant to damage.
- Wiggle the nail until the pointer moves freely and loosely around the nail. Remove the nail for now.
- About 0.4 inch (1 cm) to 0.6 inch (1.5 cm) to the right of the nail hole, glue the hair strands to the pointer. Make sure the glue has dried before starting the procedure.

PROCEDURE

- Position the pointer on the wood or Styrofoam base about three-quarters of the way down (about 7 inches [18 cm] down) with the left side of the pointer about 0.4 inch (1 cm) from the base's left side. Hammer a nail through the pointer's nail hole and securely into the base. Make sure the pointer can still turn easily.

- About 1 inch (2.5 cm) from the top of the base, in line with where the hair was glued to the pointer, hammer the other nail into the base.

- Gently pull the hair strands taut so that the pointer continues to point right, making it parallel to the ground (and perpendicular to the hairs). The strands should be perfectly vertical and the pointer should point perfectly horizontal.

- Glue the free ends of the hair strands to the top nail. *What do you think would happen if you used much shorter strands of hair? How would this affect your humidity readings?* You may need to hold the strands taut while the glue starts to set so that they retain their position. If the hairs are too long, trim the ends. Make sure the glue has dried before moving on to the next step.

- You'll next want to calibrate your hydrometer for the two humidity extremes on your scale: 0 and 100 percent. For 0 percent humidity, dry the hairs gently with a hair dryer until the strands no longer decrease in length (the pointer no longer moves). Make this location 0 percent.

- To gauge 100 percent humidity, enclose the hygrometer in the plastic box with the wet sponge or small towel. Wait at least 10 minutes. Mark the pointer location lightly, reseal the box and wait another 10 minutes. Keep doing this until the pointer location stops changing. Mark the final pointer location as 100 percent.

- For intermediate values, you could just divide the space between 0 percent and 100 percent into five equal increments and mark the locations for 20 percent, 40 percent, 60 percent, and 80 percent, then (if there is space) divide each of these in two to mark locations for 10 percent, 30 percent, 50 percent, 70 percent, and 90 percent. Your hair hygrometer may not actually be this accurate, but it can give you a rough approximation of changes in humidity.
- Use your hygrometer to measure different humidity levels around your home. *How humid does the hygrometer say it is in a room in your home? How does this compare to the humidity in the bathroom after a shower? How humid does the hygrometer say it is outside, and how does this change depending on what the weather is like?*

EXTRA

Calibrate your hair hygrometer by comparing it with another instrument, such as a psychrometer or an electric hygrometer. *How similar are the readings between your hair hygrometer and these precise instruments? If they differ, why do you think this is?*

EXTRA

Put your hygrometer outdoors in an area that is sheltered from rain. Monitor it regularly throughout the day over the course of several days. *How much does the humidity change during the course of a "normal" day? How does a storm system moving through your area affect the humidity?*

OBSERVATIONS AND RESULTS

Did your hygrometer show that it is more humid in the bathroom after a shower than in a normal room in your home? When it was used outside, did the hygrometer give a higher reading when it was more humid, such as when it rained, compared with when it was dryer?

In the bathroom after a shower, if the room is small and the door was closed the entire time, the humidity can reach around 100 percent. If calibrated correctly, the hygrometer should give a reading of about 100 percent. In other rooms, the humidity will probably be much lower than this, depending on how humid it is where you live and other factors (such as whether a humidifier or aquarium is being run). If you compared the readings of your hygrometer with another instrument that measures humidity, such as a psychrometer or an electric hygrometer, you may find that your hygrometer is 20 percent to 30 percent inaccurate, but you can recalibrate it using these devices to increase its accuracy.

As mentioned earlier, hair strands are made of protein molecules called keratins. Protein molecules are built from amino acids. In a hair strand, the keratin molecules are mainly held together by two forces: chemical cross-links between cysteines (a type of amino acid) and weaker hydrogen bonds. The keratins in hair have many such bonds, making a strand strong and flexible. The hydrogen bonds can be disrupted by water vapor in the air, with each keratin binding to water in the air instead of each other. In fact, this is what causes hair strands to regularly expand and contract in moist and dry air, respectively.

CLEANUP

Dispose of the hairs you used, or save your hygrometer to use again in the future. Return any other supplies to the place you found them.

The Big Eggshell Breakdown

PUT A DE-SHELLED EGG IN DIFFERENT LIQUIDS AND WATCH HOW IT GROWS AND SHRINKS.

Have you ever thought of a chicken egg as one big cell? Of course it is made up of many, many actual cells. But you can use it as a model to explore how different fluids get transferred from across cell membranes—and to other cells. This process is happening right now on a much smaller scale in your own body! But to see it with the naked eye we can try this macro-activity!

PROJECT TIME

10 minutes plus 3 days for observation

KEY CONCEPTS

Chemistry
Cells
Diffusion
Osmosis

BACKGROUND

Every cell of our bodies is enclosed in a plasma membrane, a complex boundary made of lots of different layers. This separates the cell contents from the surrounding world outside the cell, which is often composed of fluid. These membranes play a large role in maintaining a proper balance of contents inside the cell (homeostasis) by controlling what passes through them. One of the most important characteristics of plasma membranes is their selective permeability, which means that they allow some substances to pass freely yet restrict others. See how an egg can do this on a scale you can observe!

MATERIALS

- One egg
- A few cups or jars—large enough to hold an egg and enough liquid to submerge it
- White vinegar—enough to submerge an egg
- Large serving spoon
- Water—enough to submerge an egg
- Corn syrup—enough to submerge an egg

PREPARATION

- Place the egg in a cup or jar and then pour in enough white vinegar to completely cover it. Observe the egg. *Do you see tiny bubbles forming on the shell?*
- Cover the egg and leave it for 24 hours.

PROCEDURE

- After 24 hours, ladle the egg out of the vinegar with the big spoon. *Is the shell different? How so?*
- Dump the old vinegar and cover your now-squashy egg with fresh vinegar.

- Let the egg sit another 24 hours.
- Remove your egg and observe it again. *Has it changed again?*
- Spoon out the egg and rinse it in the sink under the tap. Be careful: because the eggshell has been dissolving, the egg membrane may be the only thing holding it together. The membrane is not as durable as the shell. *How does the egg's exterior look now?*
- Put the naked egg in a cup of water and let it rest for eight hours. *Does the de-shelled egg look the same? How has it changed?*
- Carefully remove the egg and put it in a cup of corn syrup for another eight hours. *Has the appearance of the egg changed again? How?*

OBSERVATIONS AND RESULTS

When you first place the egg in vinegar, carbon dioxide is produced as the acidic vinegar reacts with the calcium of the shell. As the egg sits in the vinegar, the shell is slowly dissolving away, producing a "de-shelled" egg. The eggshell should be totally gone by the third day of the experiment, with a blob of egg white and yolk remaining inside a thin membrane.

The de-shelled eggs are good models of human cells. After the eggshell is gone, a white membrane remains. (It's actually two membranes, but they're held tightly together.) This membrane, like those of human cells, is selectively permeable, and the entire egg can be a model for a single cell.

Substances that can pass easily through the egg's membrane will follow the principle of diffusion: They will move from the side of the membrane where they are at higher concentration to the side where they are at lower concentration. This exchange will continue until the concentrations on both sides of the membrane are equal.

Water can pass freely through a cell's membrane. (The movement of water across a membrane is a special version of diffusion called osmosis.) When an egg is soaked in a solution in which the concentration of water is higher than that inside the egg, water moves from the outside solution, across the membrane and into the egg. As a result, the egg plumps up and grows larger. This is what happened when the egg was in water.

When an egg is soaked in a solution where the concentration of water is lower than that inside the egg, however, water moves from the egg, across the membrane, and into the outside solution. As a result, the egg loses mass and may end up looking like a deflated balloon. This is what happened when the egg was in corn syrup.

The plasma membranes of our cells behave much like those of the eggs. Water, oxygen, and nutrients must pass through the plasma membrane into our cells and waste must exit. Oxygen diffuses into red blood cells in our lungs, and they transport it to our tissues, where the gas diffuses into other cells. Water in the stomach and intestines moves via osmosis into the bloodstream.

CLEANUP

Discard the egg, shells, and corn syrup into the trash. Safely dump the water you used down a drain.

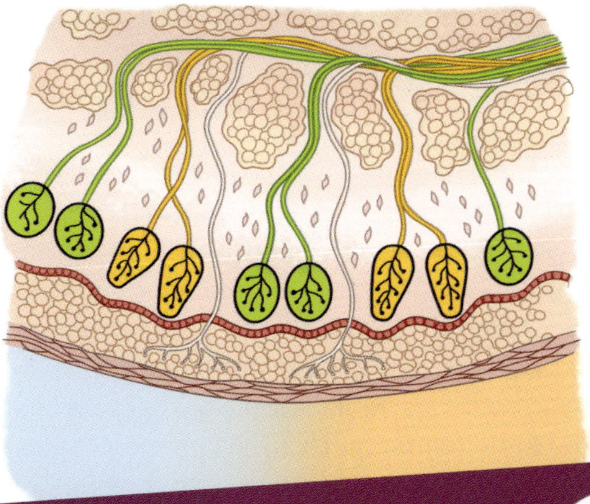

Cold or Warm, Can We Really Tell?

IS YOUR WATER HOT OR COLD? IT MIGHT DEPEND ON WHAT YOU WERE FEELING BEFORE. LEARN HOW YOU FEEL TEMPERATURE—AND HOW IT MIGHT ALL BE RELATIVE!

PROJECT TIME

15 minutes

KEY CONCEPTS

Temperature
Heat
Perception
Sensory nervous system

Have you ever tried to guess the temperature of the water in a swimming pool? On a hot day, the water might feel chilly at first, but once you're immersed in the water you don't notice its temperature as much. On a cool day, though, the pool water that is the same temperature might feel quite comfortable from the very start. Is our body equipped to tell absolute temperature? Or is it all relative? These questions might make you curious about how our bodies collect information about our environment, process it, and form our perception of the world. Do this activity, and the next time you jump in the pool on a hot summer day, you will be able to understand why you're about to feel so chilly!

BACKGROUND

Our hands—especially our fingertips—are well equipped to collect sensory information from the environment surrounding them. They contain an immense number of sensory receptors. External circumstances, such as temperature, texture, and touch prompt these receptors to produce electrical signals. The signals travel via a sensory nerve along the arm to the brain where they are processed, compared to past experiences, and finally labeled.

Each receptor is triggered by a specific stimulus. Thermoreceptors detect temperature changes. We are equipped with some thermoreceptors that are activated by cold conditions and others that are activated by heat. Warm receptors will turn up their signal rate when they feel warmth—or heat transfer into the body. Cooling—or heat transfer out of the body—results in a decreased signal rate. Cold receptors, on the other hand, increase their firing rate during cooling and decrease it during warming.

Something interesting happens when you expose receptors to a specific sensation such as heat for a long time: they start to tire out and decrease their activity, thereby you will no longer notice the sensation as much.

Could this desensitization also alter our sensitivity to what we feel next? Try this activity and find out!

MATERIALS

- Three pots, all large enough in which to submerge both hands
- Warm water (Do not make it too hot; test the water before you put your hands into the pot. If you still experience any discomfort from the warm—or cold—water, let your hands adjust to room temperature and start over using water at less extreme temperatures.)
- Room-temperature water
- Cold water (or ice cubes to add to room-temperature water)
- Towel to protect your work surface
- A clock to time yourself

PREPARATION

- Prepare a work surface that can get a little wet by laying down a towel and removing any objects that should not get wet.
- Fill one pot with very cold water. (You can also use room-temperature water and add a couple of ice cubes to cool the water in this pot.)
- Fill a second pot with room-temperature water.
- Fill a third pot with warm water. Be sure not to make the water too hot; you need to be able to comfortably have your hands in this water for a little while.

PROCEDURE

- Submerge your right hand in the pot with cold water. *How would you classify the temperature of the water—cold or very cold?*
- Put your left hand in the pot with warm water. *How does this water feel?*
- After having your hands in the pots for about a minute or two, pay attention to the temperature of the water in each pot again. *Does the cold water still feel as cold as it initially did? What about the warm water? If it feels differently, do you think the actual temperature of the water in the pots changed considerably during this short time or has your perception of the temperature changed?*
- Now, simultaneously remove your hands from the pots with ice-cold and warm water and place both hands in the pot with room-temperature water. *How would you label the temperature of the water in the pot? Does it feel hot, warm, lukewarm, cold, or very cold?* If it is hard to say, pay attention to what you would say if you felt only with your right hand and what would you say if you felt only with your left hand. *Do your hands agree or disagree about the temperature of the water?*

EXTRA

Instead of using two hands, give your index finger a warm bath and your middle finger of the same hand a cold bath. The sensory signals created by the thermoreceptor in this test run along the same sensory nerve up your arm to your brain. *Would you still be able to say one finger feels cold and the other finger feels warm? Would you still get confusing messages when after a minute, you put both fingers in water at room temperature?* Now try with a fingertip touching an ice cube and a warm cloth at the same time. *Are you still able to say that half of the tip is warm and the other half is cold? Are you still confused when you put the fingertip on a room-temperature object?*

EXTRA

In this activity, the water in the hot and cold pots are different temperatures. *What if you put your hand in contact with objects that feel cold or warm but are at the same temperature, such as a metal door knob or pot and the carpet or a wool sweater?* These objects are all at room temperature, but they appear to be different in temperature because they conduct heat differently. Let your whole hands touch these objects. *Do you still get confusing messages if, after awhile, you put your hands in contact with a third material, such as glass?*

OBSERVATIONS AND RESULTS

Did the right hand feel as though the room-temperature water was hot, whereas the left hand experienced it as chilly?

When you initially placed your right hand in the cold water, cold thermoreceptors in your hand fired creating signals that, after being processed in the brain, enabled you to label the water as "cold." As the left hand was put in hot water, warm thermoreceptors initiated signals, allowing you to identify the water in this pot as "warm."

After a while the thermoreceptors in your hands quieted down. They became desensitized and the water in the respective pots did not feel as cold or as warm anymore.

When you placed both hands in a pot of room-temperature water, however, your brain got confused. Your right hand entered with desensitized cold thermoreceptors and active warm thermoreceptors. The heat flow into the cold hand fired the warm thermoreceptors. Your brain interprets these as coming from a warm environment. You perceived the water with your right hand as warmer than it really was. A similar process happened in your left hand, which entered with desensitized warm thermoreceptors and experienced heat flow from the warm hand to the room-temperature water. Your left hand felt as though the water was colder than it really was.

As your hands perceived the water in the room-temperature pot differently, you got confused. Your brain returned conflicting information about the temperature of the water in the room-temperature pot. This experience shows that your perception of temperature is influenced by the previous environment.

CLEANUP

Safely dump the water down a drain and return the pots to their locations.

Semisolid Science
Growing Yogurt

A LIVE CULTURE CHALLENGE THAT IS SURE TO PUT YOUR SCIENTIFIC KNOWLEDGE TO THE TEST!

Have you ever wondered how yogurt is made, and why some yogurts differ from others? As most yogurt containers advertise, yogurt contains "live cultures." This means that there are living bacteria in the yogurt! These are not the harmful kind of microbes that cause you to get sick. Instead, these cultures have the amazing ability to turn plain old milk into a yummy yogurt treat. Do the bacteria affect what the resultant yogurt culture looks, feels, tastes, and smells like? In this activity, you'll find out!

PROJECT TIME

30 minutes plus 1 day of observation time

KEY CONCEPTS

Microorganisms
Bacteria
Food science
Fermentation

BACKGROUND

Bacteria, which are a type of microorganism, turn milk into yogurt. There are certain species of bacteria that are commonly used to make yogurt, and these species are good bacteria that can actually help you! If you look at the ingredients listed on the yogurt product's packaging, you can often figure out the exact species of bacteria that it contains. Some species you might find listed include: *Streptococcus thermophilus* (*S. thermophilus*); *Lactobacillus bulgaricus* (*L. bulgaricus*); *L. acidophilus; L. casei; L. rhamnosus; Bifidobacterium animalis* (*B. animalis*, or sometimes just "*Bifidus*"); and *B. bifidum*.

To turn milk into yogurt, these bacteria ferment the milk, turning the lactose sugars in the milk into lactic acid. The lactic acid is what causes the milk, as it ferments, to thicken and taste tart. Because the bacteria have partially broken down the milk already, it is thought to make yogurt easier for us to digest. Additionally, eating yogurt can help replenish the necessary populations of bacteria in your gastrointestinal tract (stomach and intestines) after they have been lost from, for example, taking antibiotics or having an upset stomach.

MATERIALS

- Six canning jars with lids, 8-ounce (236 ml) size or larger
- Large pot
- Water
- 1/2 gallon (1.8 L) of whole milk; other types of milk can be used instead too
- Candy thermometer with a range of 100 to 200°F (38 to 93°C)
- Stirring spoon
- Large double boiler (or a thick-bottomed pot) with lid
- Large pan or sink that can be plugged
- Permanent marker
- Two different types of yogurt. Try to pick types with multiple features that differ, such as one kind that is white and unsweetened (such as a Greek yogurt) and another that is artificially colored (such as by the food dye Red 40) and sweet. Use new, unopened containers.
- Two clean forks
- Measuring tablespoon
- Cooler
- Adult help and supervision with heating and handling hot liquids

PREPARATION

- Wash your hands with soap and rinse them thoroughly.
- With an adult's assistance, sterilize the canning jars, their lids, and rings. Do this by separating these pieces and putting them all in a large pot, adding about 1 inch (2.5 cm) of water, covering the pot, and boiling the water for 10 minutes. Then turn off the heat and let the jars sit, still covered, in the pot.
- Be careful when sterilizing the jars—the pot and everything inside of it will become very hot. Also be careful when handling the hot milk later in the activity.
- To successfully make yogurt, a good, sterile technique is needed. Make sure that all cookware is clean and handled properly to keep unwanted bacteria out of the yogurt cultures you are working with.

PROCEDURE

- Pour 1/2 gallon (1.8 L) of milk into the large double boiler or thick-bottomed pot.
- Heat the milk at 185 to 195°F (85 to 90.5°C), keeping the pot covered. If you use a thick-bottomed pot instead of a double boiler, stir frequently. Be careful not to let the milk boil over!
- Remove the pot from the stove and place it in a pan of clean, cool water, until the milk is close to 130°F (54°C). Alternatively, you can cool the pot in a clean, plugged sink with water.
- While the milk is cooling, prepare your jars. Carefully remove them from the pot in which they were boiled and arrange them on a clean surface. Be careful, they will be hot! Empty out any water. Do not touch the inside of the jars. Immediately put the lids and rings on each jar.

- You will be making three jars for each type of yogurt. Use the permanent marker to label three jars with the name of one of the yogurt types, and label the other three jars the other yogurt type.
- Open the first yogurt container and stir it with a clean fork. *How does the yogurt look and smell?*
- Add 1 tablespoon (15 ml) of the yogurt to each of the three appropriate jars. Put the lids back on. Thoroughly clean the measuring tablespoon.
- Open the second yogurt container and stir it with a new clean fork. *How does the yogurt look and smell?*
- Add 1 tablespoon (15 ml) of the second yogurt to each of the three remaining jars. Put the lids back on.
- Once the milk has reached 130°F (54°C), carefully pour it into the jars, filling them to about 1/2 inch (1.3 cm) from the top. Cover the jars immediately with their lids and tighten them. If you are using canning jars larger than 8 ounces (236 ml) in size, only fill them up to about 6 ounces (177 ml). (**Note:** The yogurt bacteria can be killed if exposed to temperatures above 130°F [54°C], so be careful not to add milk that is too hot!)
- Place the jars in a cooler and seal it.
- Quickly heat up about 1 gallon (3.8 L) of water until it is at 122°F (50°C).
- Add the hot water to the cooler so that the jars are surrounded, but the water is well below the lid rim.
- Put the cooler in a warm location and do not disturb it for three hours.
- After three hours, the yogurt cultures should be done if the temperature does not drop below about 100°F (38°C). Check on the jars. *How do the yogurt cultures look? Have they solidified?*

- Refrigerate the jars overnight.
- The next day, open and examine the yogurt cultures in each jar. Compare their appearance, firmness, smell, and taste to the original yogurts.
- *Did the yogurt cultures all gel? Are they firm or runny? Do they smell good or bad? How do they look, smell, and taste compared with the original yogurt that was used? How are the cultures of each type of yogurt similar or different from each other? Is this the same way the original yogurts differed from each other?*

EXTRA

In this activity, you may have focused on how the taste and color of the original yogurt affects a yogurt culture based on it, but you can explore how other aspects of the yogurt affect the resultant yogurt culture. *How do added stabilizers (such as gelatin), using organic yogurt compared with regular yogurt or other factors, like fat content, affect what a yogurt's culture is like?*

EXTRA

You can test if the amount of starter used in the yogurt culture makes a better product. *How does using more or less yogurt affect the yogurt culture? Does it take a longer or shorter time to solidify?*

EXTRA

Try testing which type of milk makes the tastiest yogurt. Try using whole, 2 percent, skim, soy, goat, or other types of milk. *How does the type of milk affect what the resultant yogurt is like? Which milk works the "best"?*

EXTRA

In this activity, you cultured the yogurt for three hours in the cooler, but varying the amount of time that the yogurt is cultured for can affect its flavor. Try culturing the yogurt in the cooler for a longer amount of time, such as seven hours. *How does increasing the culture time affect the yogurt culture? Does it look, smell, or taste different?*

OBSERVATIONS AND RESULTS

Was the yogurt in all of the jars firm and white? Did the yogurt cultures taste and smell like a very mild version of the original yogurts used?

The yogurt cultures in the jars will probably seem very similar in several ways, with some subtle differences based on the original type of yogurt used to make them. They should also be relatively firm, or firm enough so they do not slosh when tipped, and all have a similar texture. If the yogurt is not firm at all, but is actually fluid or runny, something may have gone wrong in the process and killed the bacteria—most likely the milk was too hot when added to the yogurt starters.

The yogurt cultures, however, may have small differences in taste and color based on the original yogurt used to make them. For example, if the original yogurt was really sweet, the yogurt culture should be only mildly sweet. Likewise, if the original yogurt was a bit sour (like Greek yogurt), the culture should also be a little sour. If you used one yogurt with artificial coloring (such as with Red 40) and one that was white, both resultant yogurt cultures should look white, just like the milk used to make them. If you put them side by side, however, you may notice that the culture whose original yogurt had artificial coloring is slightly off-white with a tinge of color. Overall, multiple factors affect the yogurt culture, including: the presence of some nonliving diluted ingredients from the original yogurt such as diluted Red 40 coloring, the exact process used to make the culture such as the amount of time in the cooler, and the types and amount of bacteria that were in the original yogurt.

CLEANUP

If the yogurt cultures were made correctly, you should be able to enjoy your jars of yogurt as a tasty, healthy snack! You should also be able to refrigerate the sealed yogurt for one to two weeks. The acidity of yogurt (from lactic acid) helps preserve it and prevent potentially harmful bacteria from growing.

THE SCIENTIFIC METHOD

The scientific method helps scientists—and students—gather facts to prove whether an idea is true. Using this method, scientists come up with ideas and then test those ideas by observing facts and drawing conclusions. You can use the scientific method to develop and test your own ideas!

Question: What do you want to learn? What problem needs to be solved? Be as specific as possible.
Research: Learn more about your topic and refine your question.
Hypothesis: Form an educated guess about what you think will answer your question. This allows you to make a prediction you can test.
Experiment: Create a test to learn if your hypothesis is correct. Limit the number of variables, or elements of the experiment that could change.
Analysis: Record your observations about the progress and results of your experiment. Then analyze your data to understand what it means.
Conclusion: Review all your data. Did the results of the experiment match the prediction? If so, your hypothesis was correct. If not, your hypothesis may need to be changed.

GLOSSARY

adaptation: Changing or making adjustments to better fit one's surroundings.

allele: Genetic sequence that a person receives from their parents at conception.

biogeochemical: Having to do with matter being moved between living systems and the environment.

biotechnology: Technology that utilizes biological systems and/or living organisms to develop and create new products.

defense mechanism: An organism's defensive reaction.

diffusion: The process of a substance moving from an area of higher concentration to an area of lower concentration.

evolution: The process of living organisms growing and changing throughout the course of the Earth's timeline.

gradient: An increase or decrease in temperature, pressure, or concentration passing from one point to another.

insulator: A substance that blocks the passage of heat or sound.

metabolism: Chemical changes that take place in a cell or an organism.

microbe: A tiny living thing that causes disease or fermentation.

microorganism: A tiny living thing, such as a bacterium.

osmosis: The movement of water from an area with a lower concentration of a solvent to an area with a higher concentration of the solvent in order to make the solution equal in concentration.

photosynthesis: The process by which green plants and some protists use light to turn water and carbon dioxide into food.

regulate: To control and maintain the rate or speed of a machine or process.

synthetic: Made by chemical products to imitate a natural product.

thermal: Having to do with heat.

thermoreceptor: Nerve ending that reacts to changes in temperature.

ADDITIONAL RESOURCES

Books

Dickinson, Michelle. *Kitchen Cabinet Science Projects: Fifty Amazing Science Experiments to Make with Everyday Ingredients.* New York, NY: Penguin Workshop, an imprint of Penguin Random House LLC, 2021.

Heinecke, Liz Lee. *Sheet Pan Science: 25 Fun, Simple Science Experiments for the Kitchen Table.* Beverly, MA: Quarry Books, 2022.

Olson, Elsie. *10-Minute Kitchen Science Projects.* North Mankato, MN: Dabble Lab/Capstone Press, a Capstone imprint, 2022.

Websites

Discovery Education
sciencefaircentral.com

Exploratorium
www.exploratorium.edu/explore/activities

Science Buddies
www.sciencebuddies.org/science-fair-projects/project-ideas/list

Science Fun for Everyone!
www.sciencefun.org/?s=science+fair

Videos

Kitchen Science | DIY Science Time
https://ny.pbslearningmedia.org/resource/b4ad5b8d-d11f-4bde-8aac-2254648fbf1a/kitchen-science-video-diy-science-time/

The Common Genetic Code
https://ny.pbslearningmedia.org/resource/tdc02.sci.life.gen.comgencode/the-common-genetic-code/

Publisher's note to educators and parents: Our editors have carefully reviewed these websites to ensure that they are suitable for students. Many websites change frequently, however, and we cannot guarantee that a site's future contents will continue to meet our high standards of quality and educational value. Be advised that students should be closely supervised whenever they access the internet.

INDEX

B

bacteria, 11, 23, 26, 27, 54, 55, 56, 57, 59
birds, 16, 17, 18

C

camouflage, 15, 16, 17, 18
carbon dioxide, 10, 11, 12, 13, 14, 47
cells, 33, 34, 35, 37, 40, 45, 46, 47, 48
clothes, 6, 9, 21, 22

D

DNA, 4, 29, 33, 34, 36, 37

E

ecosystem, 20, 21
extraction, 4, 33, 34, 35, 36

F

fat, 6, 7, 8, 9, 58
feathers, 6, 7, 9
fermentation, 4, 10, 11, 12, 14, 54
food, 6, 11, 14, 54, 55
fungus, 11

G

genetics, 28, 32
genome, 33, 34

H

habitat, 16, 17, 18, 19
hair, 30, 31, 32, 39, 40, 41, 42, 43, 44
heat transfer, 5, 8, 50
hypothesis, 61

I

inheritance, 28, 32

K

katydids, 16, 18
keratin, 40, 44

M

mammals, **6**
Mendel, Gregor, 29

N

nutrients, 20, 21, 26, 48

O

octopuses, 16
oxygen, 20, 21, 26, 48

P

perception, 49, 52, 53
plants, 11, 29
plasma membrane, 46, 48
polar bears, 6
predators, 15, 16, 17
prey, 15, 16, 17, 18, 19

S

science fair ideas, 8, 13, 18, 25, 26, 36
sensory receptors, 50
soil, 20, 21, 23, 26
sugar, 11, 12, 13, 14, 55

T

temperature, 5, 6, 7, 8, 12, 13, 22, 24, 25, 26, 49, 50, 51, 52, 53, 57
traits, 28, 29, 31, 32

Y

yeast, 4, 10, 11, 12, 13, 14